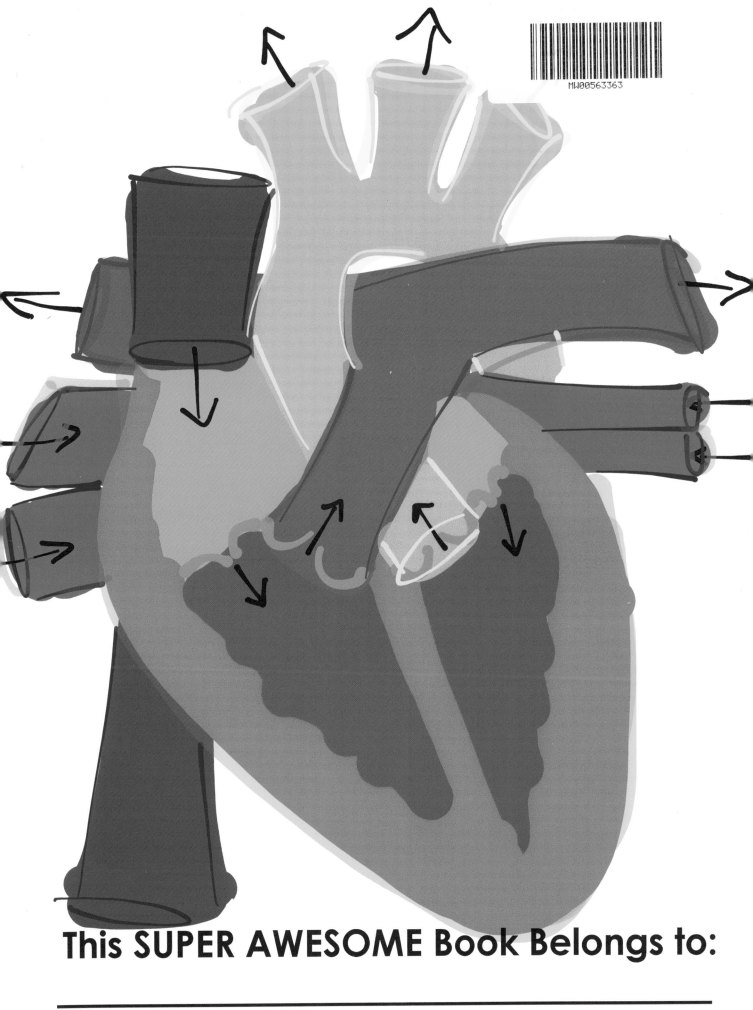

This **SUPER AWESOME** Book Belongs to:

_____

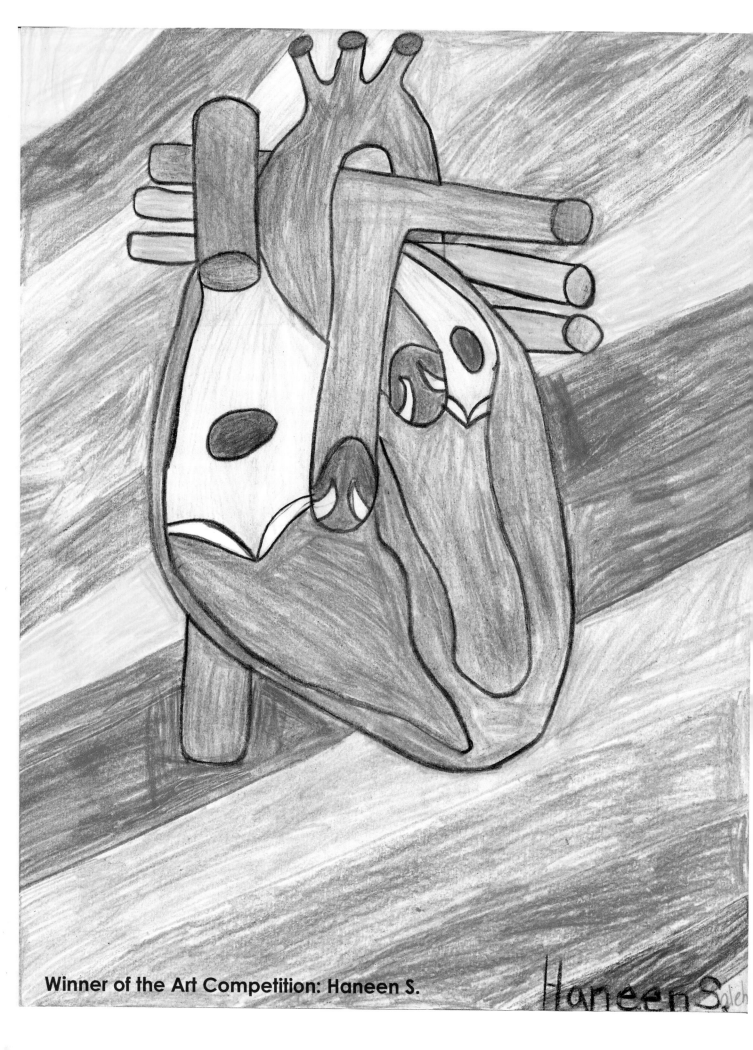

Winner of the Art Competition: Haneen S.

Cardiology FOR KIDS

...and Adults too!

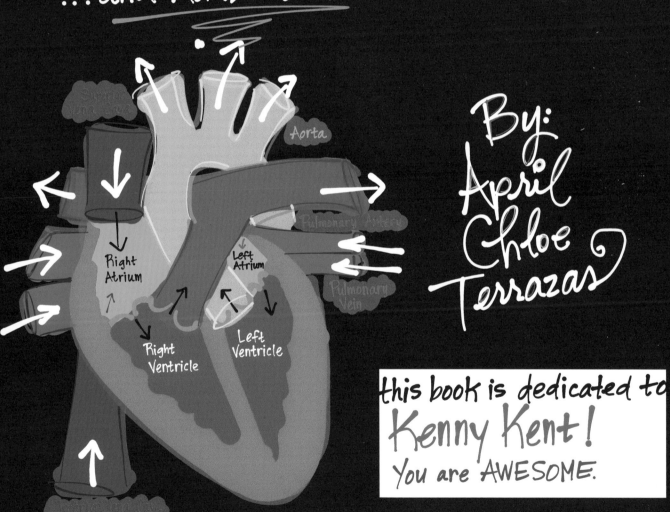

By: April Chloe Terrazas

this book is dedicated to **Kenny Kent!** You are AWESOME.

Cardiology FOR KIDS ...and Adults too!   By: April Chloe Terrazas, BS University of Texas at Austin.

Copyright © 2014 Crazy Brainz, LLC

Visit us on the web! www.Crazy-Brainz.com

Cover design, illustrations and text by: April Chloe Terrazas

Cross your right hand over your chest. That is where your heart is located.

Make a fist with your hand. That is about the size of your heart.

Take two fingers and touch the side of your neck.

Can you feel your heartbeat?

When you finish this book, you will know what every colored part of the heart does!

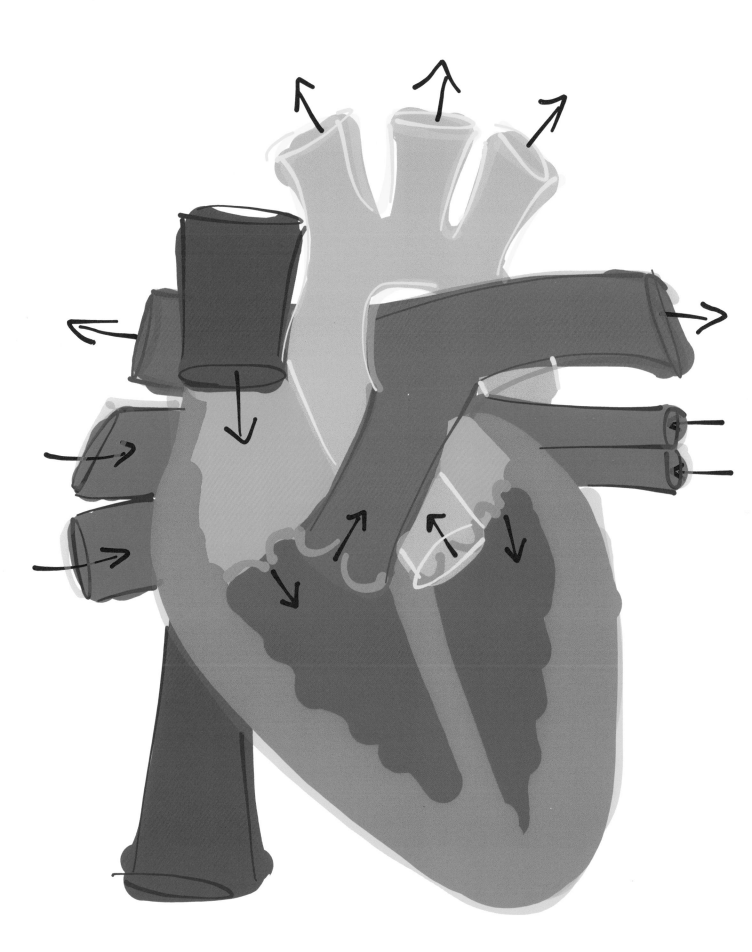

The main station in the Circulatory System Railroad is the heart.

The "tracks" on the railroad are:

Arteries: AR-TER-EEZ

Arterioles: AR-TER-EE-OLZ

Arteries are large "tracks" (blood vessels) that leave directly from the main station, the heart. Arterioles are smaller "tracks" that lead from the larger arteries to an even smaller blood vessel.

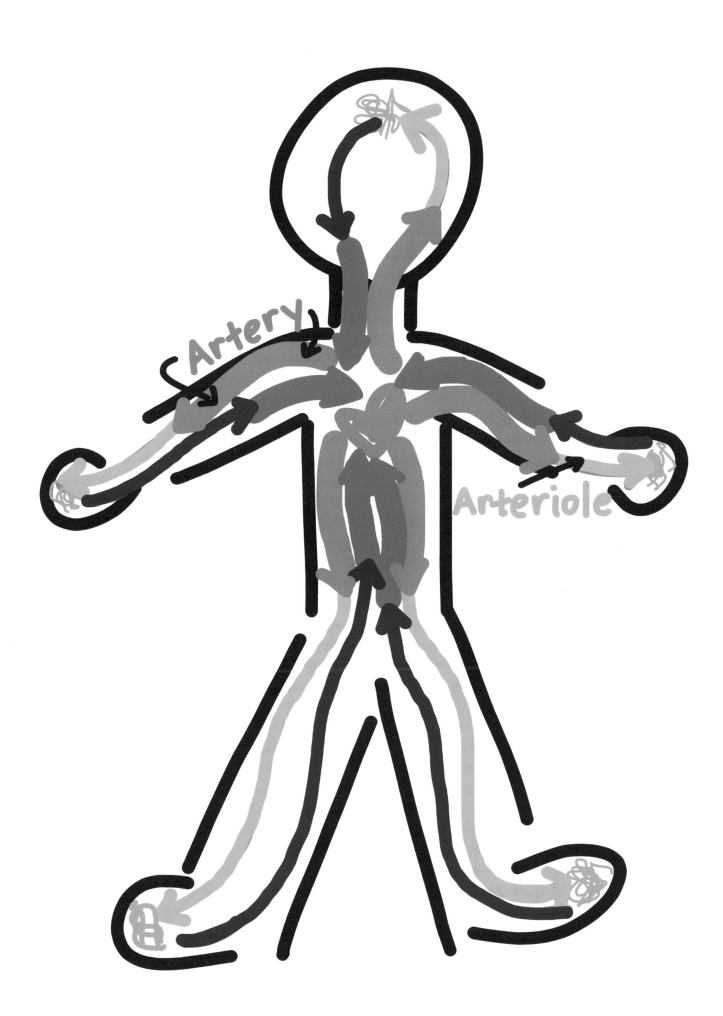

Capillaries: CAP-IL-AIR-EEZ

Venules: VIN-ULZ

Veins: VAYNZ

Arterioles lead to capillaries. Capillaries are very small!

Venules connect on the other side to the tiny capillaries.

Veins are larger "tracks," like arteries. Veins bring blood back to the heart to get more oxygen.

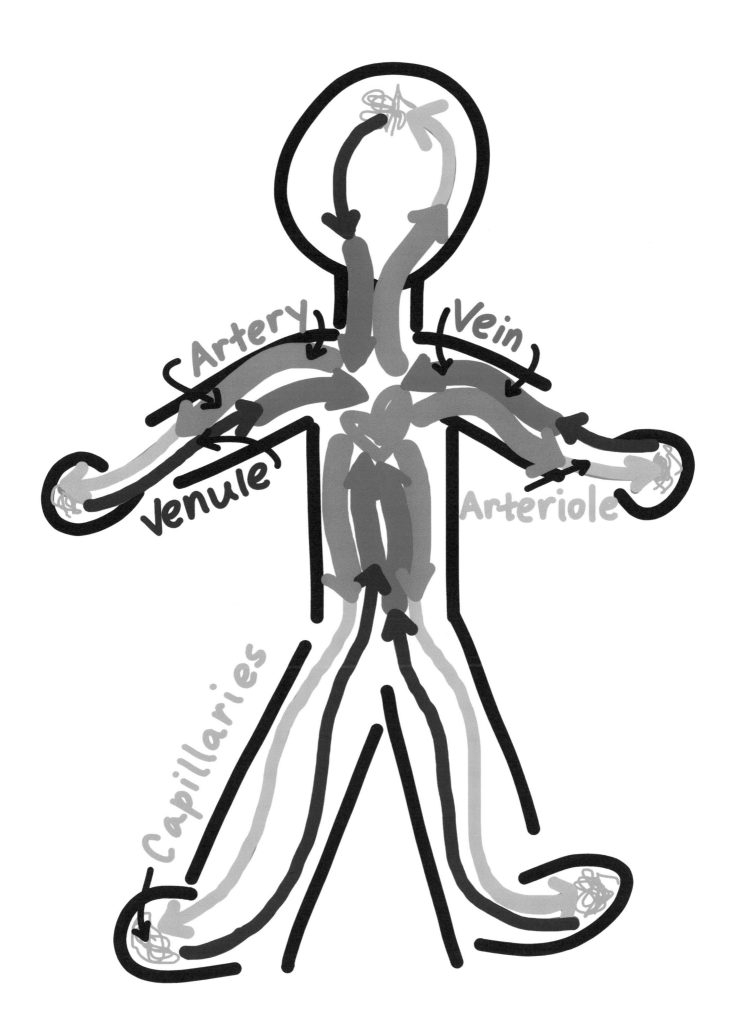

What is the main station of the circulatory system railroad?

What are the names of the "tracks" on the circulatory system railroad?

Where is your heart located?

What is the size of your heart?

Where can you feel your heartbeat?

The right side of the heart receives blood from the body and sends it to the lungs.

When this blood reaches the lungs, it exchanges CO_2 (carbon dioxide) for O_2 (oxygen).

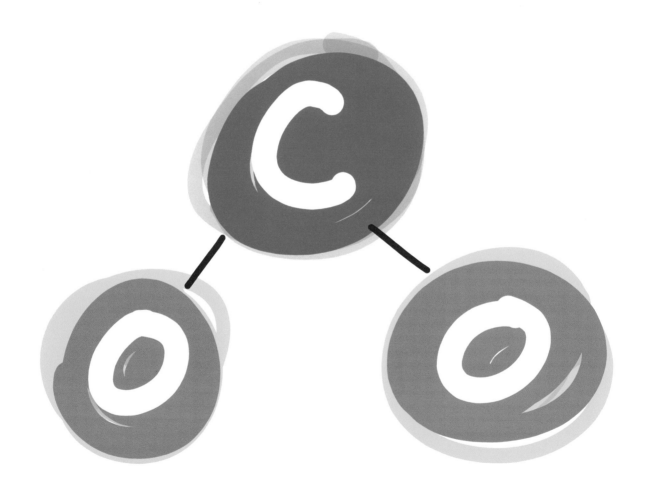

What is the name of this molecule?

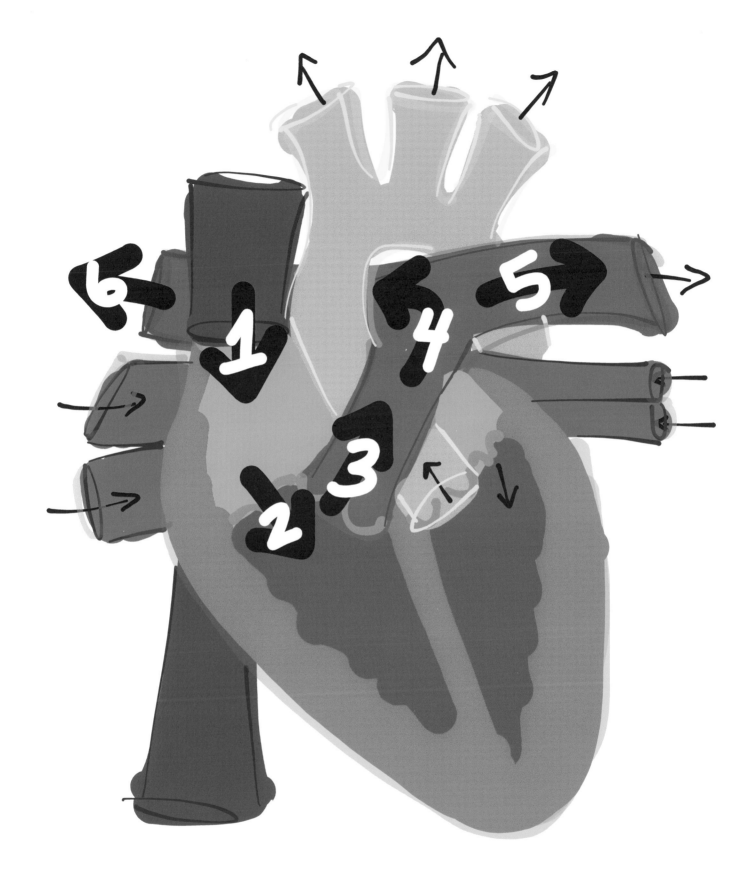

#5 and #6 are heading toward the lungs.

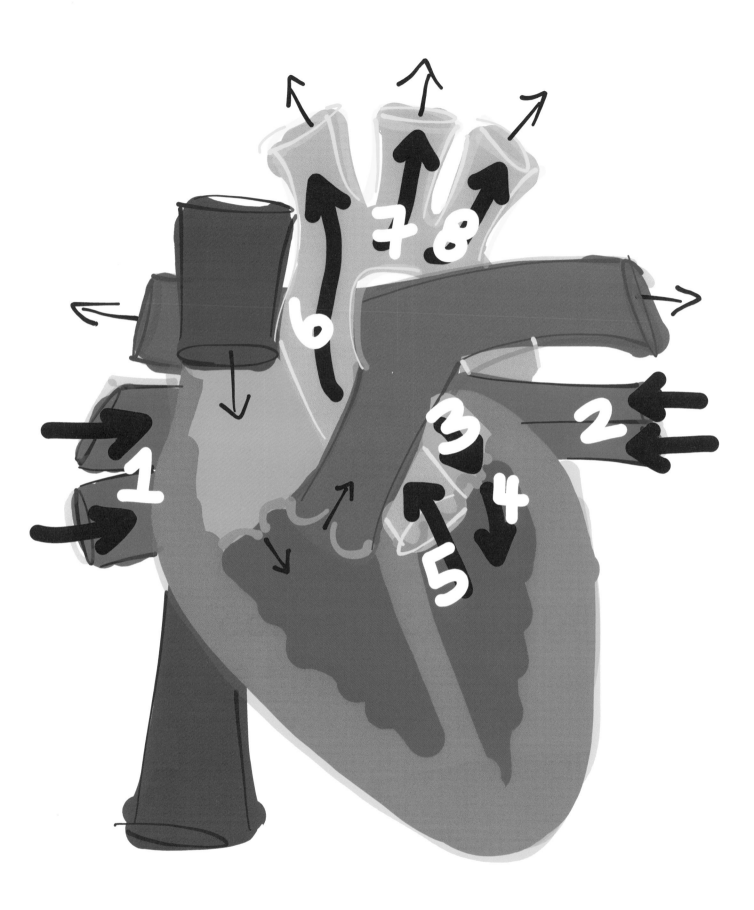

The blood returning to the heart from the lungs (through #1 and #2) now has O_2 (oxygen).

The newly oxygenated blood coming from the lungs enters the left side of the heart (#3, #4 and #5).

It is then pumped back out, delivering O_2 (oxygen) to all of the cells in the body.

#6, #7 and #8 are heading out all over the body.

The right side of the heart receives blood from the _____?

The right side of the heart sends blood to the _____?

What does the blood exchange when it reaches the lungs?

What is CO_2?

What is O_2?

The left side of the heart receives blood from the _____?

This blood is now carrying _____ that it received from the lungs.

The blood leaves the left side of the heart and travels all over the _____.

Well done!

The heart has four main areas called chambers.

The 2 chambers on top are called atria (an individual chamber is called an atrium).

There is a left and a right atrium.

Atria

Sound it Out
1. A
2. TREE
3. UH

Atrium

Sound it Out
1. A
2. TREE
3. UM

The atria receive blood from the body and from the lungs.
Blood always enters the atria first.

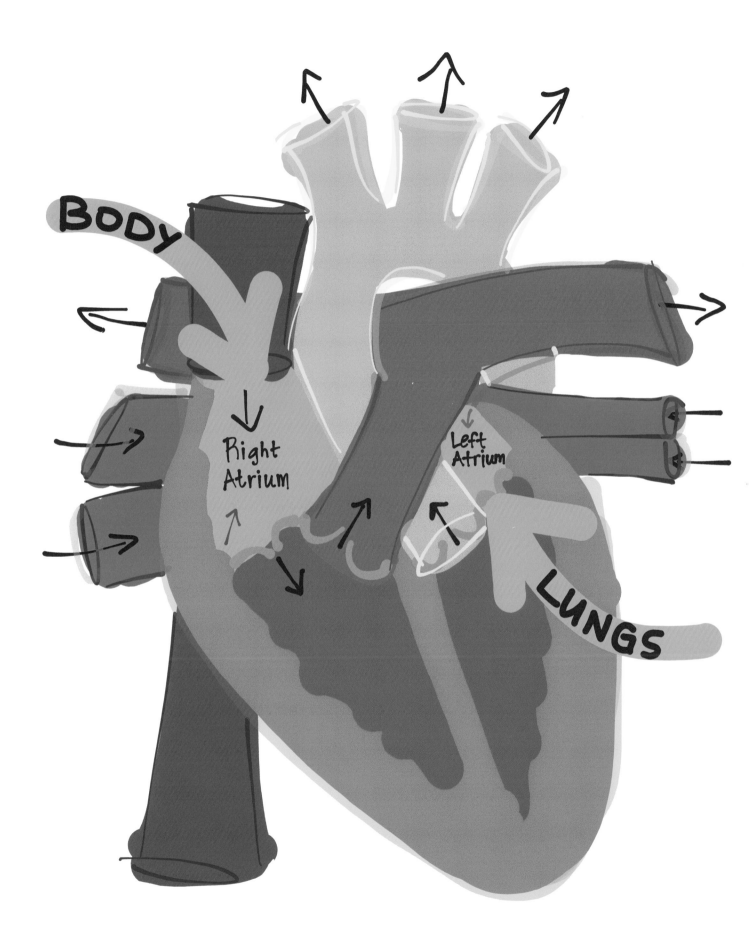

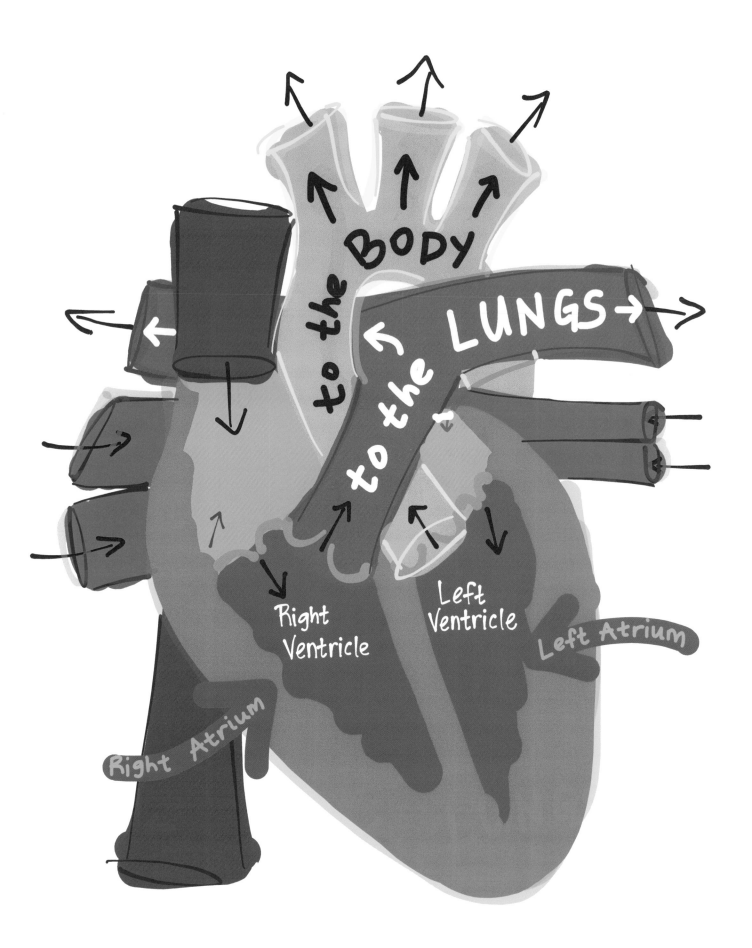

The 2 chambers on the bottom are called ventricles.

Ventricles

Sound it Out
1. VIN
2. TRE
3. KLZ

Ventricles receive blood from the atria and then move blood out of the heart, to the body and lungs.

You have a left ventricle and a right ventricle.

The right atrium receives oxygen-poor blood from the body through the superior and inferior vena cava and passes it along to the right ventricle. The vena cava is a large vein.

Superior (SOO-PEER-EE-OR) = top.
Inferior (IN-FEER-EE-OR) = bottom.

The right ventricle then pushes the blood out of the heart and toward the lungs where it picks up oxygen.

Vena Cava

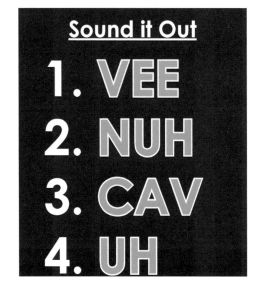

Sound it Out
1. VEE
2. NUH
3. CAV
4. UH

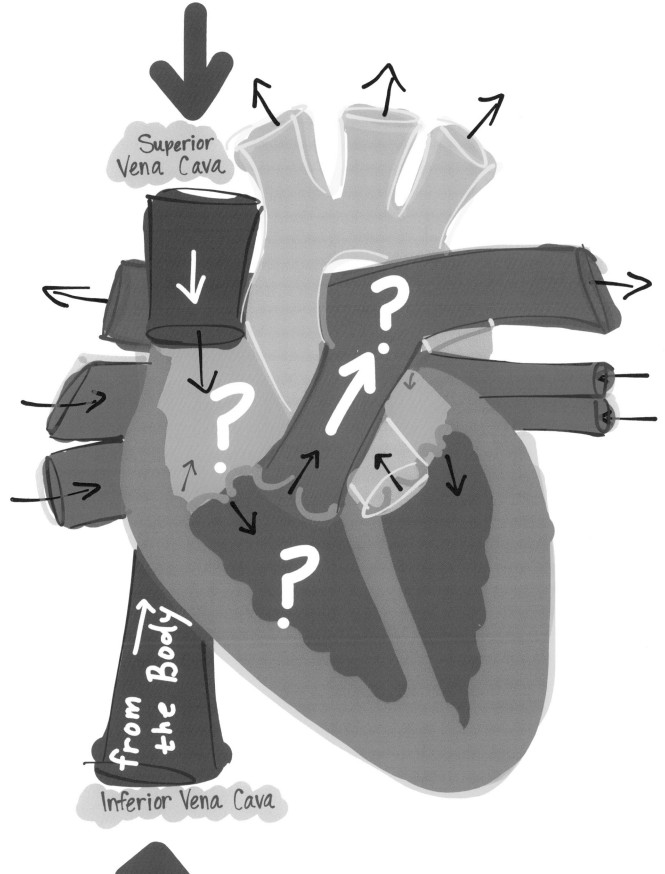

Superior
Vena Cava

Inferior Vena Cava

from the Body

How many chambers
are in the heart?

The chambers in the top of the
heart are called _____.

The chambers in the bottom of
the heart are called _____.

Which chamber does
blood enter first?

What does SUPERIOR mean?

What does INFERIOR mean?

What is the name of the large vein that brings blood to the right atrium?

Fantastic!

The blood leaves the right ventricle and travels through the pulmonary artery toward the lungs.

Pulmonary is a word you use when speaking about the lungs.

The pulmonary artery goes to the lungs.

Pulmonary

Sound it Out
1. PUL
2. MUH
3. NER
4. EE

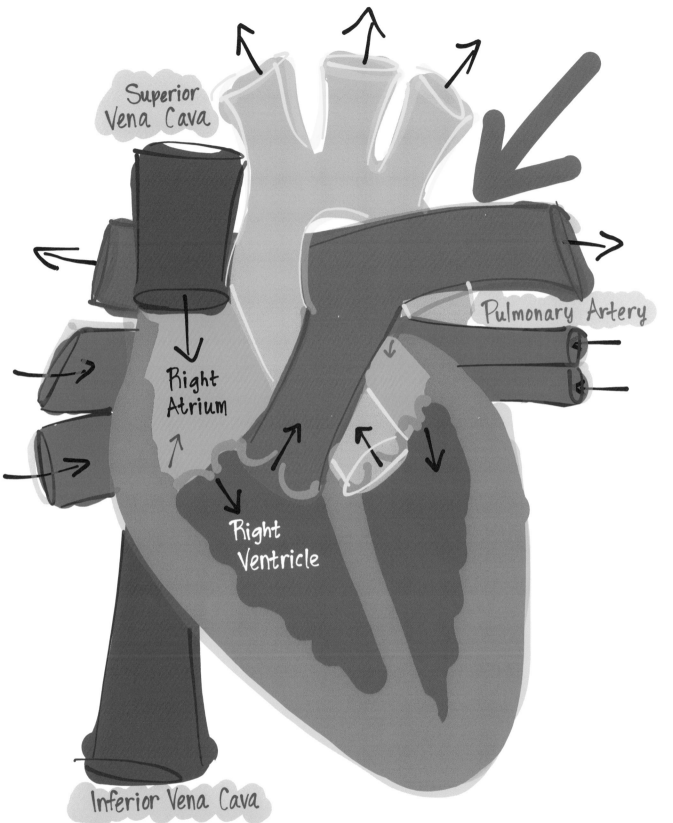

Superior Vena Cava

Pulmonary Artery

Right Atrium

Right Ventricle

Inferior Vena Cava

The blood receives O_2 (oxygen) from the lungs and travels back to the heart.

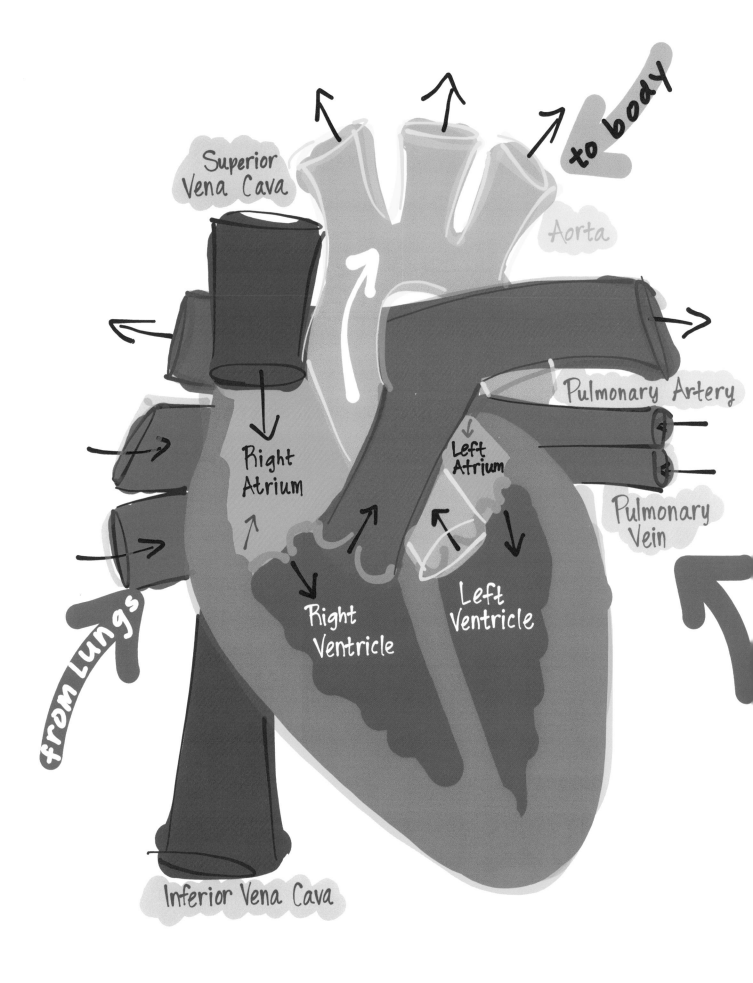

After getting oxygen in the lungs, the blood comes back through the pulmonary vein and into the left atrium of the heart.

The blood moves to the left ventricle where it is pushed out to the aorta and to the body. The aorta is the main artery of the body.

Aorta

Sound it Out

1. A
2. OR
3. TUH

What does the word "pulmonary" refer to?

Blood follows the path from the superior and inferior _____, to the right _____, then to the right _____.

When the blood leaves the right ventricle, what does it travel through? *(hint: it's pink!)*

What does blood pick up in the lungs?

After blood gets _____ from the lungs, it travels back to the heart through the _____ ____.

What is the name of the main artery of the body?
(hint: it's orange!)

Very good!

Did you know that your heart has doors just like a house?

The blood stops in the right atrium "room", goes through the door (tricuspid valve) to the right ventricle "room", stops, goes through the door (pulmonary valve) to the pulmonary artery.

The heartbeat sound is made by the closing of the valves.

Tricuspid

Sound it Out

1. TRI
2. KUS
3. PID

Mitral

Sound it Out

1. MI
2. TRUL

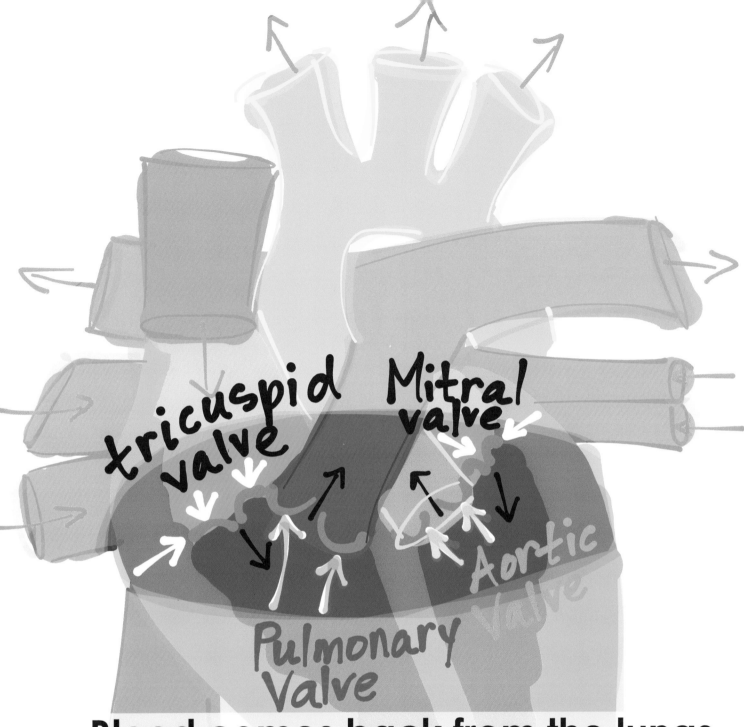

Blood comes back from the lungs into the left atrium "room," stops, goes through the door (mitral valve) to the left ventricle "room," stops, and goes through the door (aortic valve) to the aorta.

Heart contraction starts at the SA node (sinoatrial node). It is located in the right atrium.

This sends a signal (in yellow) to the AV node (atrioventricular node). It is located between the right atrium and right ventricle.

The AV node continues the signal down the purkinje fibers which causes the ventricles to contract, moving blood to the body and lungs!

When the heart relaxes, it fills with blood again.

Purkinje: PUR-KIN-JEE

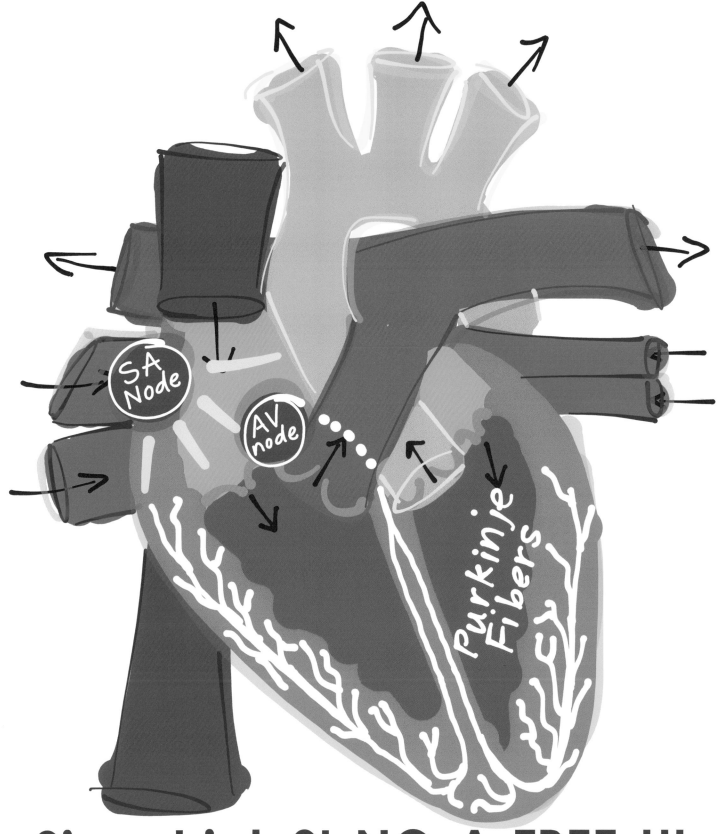

Sinoatrial: SI-NO-A-TREE-UL

Atrioventricular:
A-TREE-O-VIN-TRIK-U-LER

What are the "rooms" of the heart?

What are the names of the valves that go from the atria to the ventricles? *(hint: they are grey!)*

What are the names of the valves that go from the ventricles leading out of the heart? *(hint: one is pink, one is orange)*

Where does the heartbeat start?

The signal travels to the _____ _____, located between the atrium and ventricle.

What is the name of the fibers that spread around the heart, causing it to contract?

What makes the heartbeat sound?

YOU ARE AMAZING!

Your heart is located in your chest and is about the size of your fist. You can feel your heartbeat by touching the side of your neck.

The "tracks" on the circulatory system railroad are: arteries, arterioles, capillaries, venules and veins.
The "tracks" transport O_2 and CO_2.

The right side of the heart receives blood from the superior and inferior vena cava. It then travels to the right atrium, through the tricuspid valve, into the right ventricle, through the pulmonary valve and out the pulmonary artery toward the lungs.

Oxygen-rich blood returns from the lungs through the pulmonary vein. It then travels to the left atrium, through the mitral valve, into the left ventricle, through the aortic valve and out the aorta toward the body.

Heart contraction is generated by the SA (sinoatrial) node. The signal then goes to the AV (atrioventricular) node. The AV node sends the signal to the purkinje fibers which cause the ventricles to contract, moving blood out of the heart.

When the heart relaxes, blood moves back into the heart. The heartbeat sound is made by the closing of the valves.

NEW VOCABU

Arteries

Arterioles

Capillaries

Venules

Veins

Atria

Atrium

Ventricle

Superior

Inferior

Vena Cava

LARY!

Pulmonary

Aorta

Tricuspid Valve

Mitral Valve

Pulmonary Valve

Aortic Valve

SA (sinoatrial) node

AV (atrioventricular) node

purkinje fibers

You are now a Cardiology expert!

Draw your heart, with the atria, ventricles, aorta, pulmonary artery, pulmonary vein, valves and nodes!

CPSIA information can be obtained at www.ICGtesting.com
Printed in the USA
BVIW12n2055020118
503998BV00011B/115